Best Wishes To the
Seaside Signal,
my old alma mater!

Ky Ward Jennings

May 25, 2010

Contents

A WINDOW TO THE SEA

Astoria, Columbia River Communities
Clatsop Plains, Gearhart
Seaside And Cannon Beach

From A River To The Sea

BY
Ky Weed Jennings

ISBN 978-09764219-1-7

Library of Congress
Categories 900, 770

Front Cover Photo
Lisa K. Johnston

Back Cover Photo
Angela Baumgartner

Edited by
Judy Wilson
Jennifer Tythcott

WPP
Winchester Place Publishing
Portland, Oregon

Regal Printing, Inc. HK

Introduction

Living in what was a small resort town of about 3800 people, knowing everyone, their families, and the houses they lived in, and knowing the people who came for the summers, was the most security anyone could ask for. We were all pretty much in the same boat in the times we grew up after WWII. Our parents were optimistic after the war, and there was a sense of togetherness hard to describe to people who are not of our generation. No one was any better than anyone else or worse; we all had only a front and back door to our small houses and neither of them were ever locked.

This small history of the area in pictures and stories is a glimpse of our lives then. It's about how we lived and who some of our families were. Letters and stories have come in from people who grew up in the area or who came to visit; people with the same time and place in common who all share a special bond to these small seaside communities where the river meets the sea.

My father flew a little airplane. It was 1949 and from my window of the plane, I saw a merry go round, a beach, a great blue ocean, and this was my view through the window of the airplane. My mother wouldn't fly with him but I volunteered and it was an adventure every time up in the sky. It looked like a village of houses and buildings below to me, and it was like looking at a storybook or playhouse. We would fly between the mountains that surrounded one side and the clouds were like cotton candy, only white instead of pink. He'd tip the wings of the plane and I'd look out to a village of toys from the sky. We'd fly over Seaside and Gearhart, over by Ilwaco where there were lighthouses on cliffs. Then we'd be back over Astoria, where there were slowly floating ferryboats on the Columbia, and hundreds of small fishing boats. Then down along the ocean, to Cannon Beach that had rocks sticking out of the sand that looked like mountains. People say we "lived in a bubble" during this time at the coast, and this is my view from inside the bubble.

All Roads Lead to the Sea

"Go down to the sea
Put your ears to the ground
And open your heart to the waves
And the sounds that they make
As they break along the shore
Never more to pass this way again."

Once in awhile, in a moment of time, you can fall in love with a place. It's somewhere you like to be, and you know you are home. This is a place like that, a crossroads, where the river meets the sea.

There are no words to describe the beauty of the ocean in your front yard. The crystals of sand in the air, the breeze in your face, the sounds of the ocean, the smell of fresh sheets hanging on the clothesline, all this, a gift from the sea.

Astoria

the New York of the Pacific

Celebrating their Norse Heritage, Scandinavians in Astoria sponsored a week-long Centennial celebration in April of 1911.

Queen Beatrice, the first Regatta Queen

Astoria Centennial

The only transportation from one side of the Columbia River to the other, was by canoe, boat or ferry. This was the old territory that eventually became Washington and Oregon. The Chinooks and the Clatsops lived along the banks of the Columbia. They knew of the violent rushing currents that collided when the Pacific and the Columbia crashed together and often would guide ships coming in. Astoria had been inhabited by Native American Indians until the immigrants arrived from Finland, Sweden, Norway, Russia, China, Japan, and French Canada. Astoria was considered to be the New York of the West with an entry point surrounded by the Pacific Ocean, the Columbia River, bays and streams. There were communities of people from the old country with their own ways who kept much of their culture, their language, and determination and became the melting pot of the Pacific Northwest. They came with their few belongings and their strong spirits to Astoria and surrounding areas on both sides of the Columbia.

Commercial Street

Life Saving Crew, Ilwaco

Around The Columbia

A Clatsop Indian village is said to have been located on Youngs Bay, directly south of Astoria High School. Astoria was settled on a peninsula rising steeply from the Columbia River's edge. The city was first platted in 1844 and those who settled in Astoria constructed hotels, restaurants, banks and sawmills on wood pilings over a bay just below Ft. George.

Wooden streets, sidewalks and piles needed constant maintenance, and the potential for fire was high. Public works projects constructed a "seawall" along the river's edge to block tidewater. The commercial district was raised as much as four feet to provide room for basements. Concrete retaining walls were constructed to support a new street system. Two fires took their toll. In 1883, a fire burned the eastern portion of the town.

JOHN WILSON

A Finnish Immigrant in Astoria

Rescue of the American Steam Schooner, Washington, November 17, 1911

John Wilson was part of the crew of the tug, Tatoosh, that rescued the American steam schooner, Washington, November 17, 1911, off Peacock Spit. The crew members all received medals for bravery for their efforts to rescue not only the passengers, but the steamer as well.

On Nov. 17, 1911 the steam schooner, Washington, heavily laden with lumber, was fast aground on Peacock Spit. For twenty hours the steamer was buffeted by nasty seas. It would have been extremely dangerous to attempt to rescue the passengers, but Captain C. T. Bailey of the Tug, Tatoosh, wasn't content for just that. He planned to save all or nothing and it proved to be the feat of the decade.

Excerpted from "Pacific Graveyard"

by James A. Gibbs

One of John Wilson's four sons, Fred Wilson, owner of Wilson Studios, became a well-known Astoria photographer, honored in the United States for his portraiture.

An Astoria Community Band includes Fred Wilson, Ed Wilson and Dick Wilson.

The Imperial Restaurant Astoria

Megler Landing

SEINING
TROLLING
GIL NETTING
FISHING

Above, Frank Alto, on the Tyee, which he owned with his brother-in-law, Bill Rassina. They fished out of the West End Mooring Basin in Astoria. The boat was converted from fishing for soup fin sharks for vitamins in WWII, to trolling for salmon and tuna. They started fishing in 1946 for salmon and tuna from Newport to the Columbia River. The only electronics on the boat was a radio direction finder which they used to home in on KAST radio when returning from off shore. They had no radios, depth finder, radar, or GPS. They sometimes had to run day and night for three days to reach Astoria from the off shore tuna grounds. The Tyee had a gasoline motor with 800 gallons under the decks. All of the boats fished alone and were unable to keep in touch with each other.

ASTORIA

Until the railroad arrived in 1898 all transportation to and from Astoria was by water. Mass transportation within the city was provided by horse-drawn or electric streetcar. Then in 1915, the Columbia River Highway connected Astoria to Portland. Travel to Seaside was by construction of highway bridges across Youngs Bay and the Lewis and Clark River. Railroad trestles constructed by the Astoria & Columbia River Railroad in 1896 remain exposed and in use from the foot of Columbia Avenue to 2nd Street, from the foot of 6th to 11th streets, at the breach of the former Clatsop Mill and three 100 foot plus sections from 42nd to 51st streets. Piles from the Astoria-Seaside Railway were driven in 1896, two years before Astoria had rail service to Portland. The piles were removed from the bay in the late 1980s. The Beltline Railroad was constructed in 1919 on the north shore of Youngs Bay.

In the spring of 1811, a party of 33 men arrived at the mouth of the Columbia River on the bark "Tonquin". Inspired by the discoveries of Lewis and Clark, John Jacob Astor commissioned these men to establish a fur-trading post for the Pacific Fur Company. A site was selected for the post on what is now the southeastern edge of the Astoria business district. While not constructed for military defense, the post did provide some feeling of security from the indigenous people. In 1813, the post was sold to the Northwest Fur Company, a British venture, and renamed Fort George. Palisades and garrisons were added at that time. Although the Treaty of Ghent was signed in 1814, a transfer of the fort to the United States was made in 1818. Astoria remained occupied by the British until the 49th north parallel was established as the United States' northern boundary in 1846.

During the Civil War, Fort Stevens (located 10 miles outside Astoria), Fort Columbia and Fort Canby in Washington were established to protect the mouth of the Columbia River. It was feared that the British might join forces with the Confederate army and assault the area. These forts were used through World War II. Another product of the Second World War was the development of Tongue Point as a Naval Air Station. The military constructed deep water wharfs, hangars, a hospital, officer housing, dormitories and multiple-family houses.

White Collar Line's Sternwheeler, Bailey Gatzert, was originally built in 1890 and rebuilt in 1907, to be the finest, most elegant steamer on the Columbia River.

A side-wheeler was constructed by the Oregon Railway and Navigation Company in Portland, Oregon in 1888. The 230-foot long vessel was noted as the fastest steam-driven paddle wheeler to ever run on the Columbia River. In 1901 it was remodeled and enlarged as one of the most luxurious riverboats in the Northwest. The vessel was condemned in 1916 and it then served as barracks for workers at the nearby Wilson Shipyard. In 1925, it was beached then stripped and burned by salvagers.

Great Northern
Astoria to San Francisco

"The San Francisco Fair was in 1915. Two big ships, the "Great Northern" and the "Northern Pacific", ran from Flavel (North Warrenton) to San Francisco. I had the privilege of going aboard the "Great Northern" the first day she docked. They could make the trip to San Francisco in 26 hours. Later on after the United States got into World War I, these two ships hauled the most troops to France."

by Clarence Sigurdson, from "Raised by the Sea"

Bumble Bee Cannery

Astoria's major industries included canneries, sawmills, and shipbuilding. Astoria's first cannery was the third in Oregon and was incorporated in 1873. From then, canneries multiplied quickly and within two years, Astoria operated eleven canneries. Within twenty years the number of canneries would double. Seven canneries combined their plants and equipment in 1899 to form the Columbia River Packers Association. That same year, 200 fishermen organized their own co-op cannery, Union Fishermen's Co-op Packing Co.

John Jacob Astor Park was constructed in 1926 by the Great Northern Railroad and Vincent Astor. The 125 foot high monument commemorates area history from the "Forest Primeval" to the coming of the railroad in 1898.

The USS Oregon in Astoria

The USS OREGON was one of the first true battleships created in 1890 to update the U.S. Navy so that it could defend against the imperial world powers, as well as by the nations of South America. It was strictly for coastal defense, a sea-going coastline battleship. The race of the USS OREGON around South America and through the stormy Strait of Magellan to join the Atlantic Fleet, was closely followed by the American public. During the Battle of Santiago, in 1900, the OREGON carried troops to China during the Boxer Rebellion.

Picture of the voyage of the Oregon from San Francisco to Santiago, April, 1898.

LIBERTY THEATRE

Portland architects, Bennes and Herzog, designed the Mediterranean inspired theater and building in 1925. One of Astoria's finest buildings, the theater featured small vaudeville acts, silent films and a magnificent Wurlitzer organ. It remains today as one of the jewels of Astoria's business district. The building is listed on the National Register of Historic Places.

Youngs Bay Mill

"She loved to ride those ferry boats,
She spoke of life and talked of hopes,
to sail the finest ships afloat,
to chase her rainbow to the end."

The Astoria clowns

by Cindy and Jeff Daly

In 1955 some Astoria merchants, organized under the Chamber of Commerce Retail Merchants Committee, entered the Lewis & Clark Sesquicentennial Regatta parade as clowns. The clowns had so much fun and received so much attention in the parade that their performance was repeated in the Astoria Regatta Parade in 1956 and 1957 with a growing number of props.

In 1958 the group organized as a non-profit corporation, separate from the chamber. By then it had 15 members. The 1948 Chrysler was painted bright yellow and covered with the bridge slogan "Come on Oregon Let's Build the Bridge." That year the clowns paraded in Castle Rock and Longview, Washington and Florence, Oregon in addition to the annual Regatta. In 1959 the schedule included the Seattle Sea Fair and Portland Rose Festival Twilight Parade along with a number of other out-of-town appearances.

In 1960, the clowns were the first group ever allowed in the Rose Festival parade not covered in roses. So many parades were being scheduled that in 1965 membership was increased to 20, going to 25 a few years later. Soon membership was expanded to 30 active clowns and each new member is elected by unanimous vote. The members try to remain anonymous and agree to attend at least six out-of-town parades a year.

In 1966 when the Astoria Bridge finally opened, the first car to cross after the Mayor's car was a 1951 Dodge seven passenger former funeral car now painted rescue orange and filled with 15 clowns. The first day was free and the next morning when tolls were charged, the Astoria Clowns, driven by John Gustafson, paid the first toll of $1.50. Vic Watson, Bob Lovell, Chuck DeFoe, Vern Mogenson, Bob Graves, Bob Wooldridge, Wes Shaner, Chuck Gustafson, Jack Daly, Jean Halleaux, Butch Mashke and Danny Jones went along for the ride. When the tolls were removed in 1993, the clowns celebrated in their present parade car, a retired Cadillac hearse, and were the first to cross without a toll.

Once the bridge was completed, the clowns began to promote Astoria history and tourist appeal. After ten years of active membership a clown can apply for life membership. Life members include Jack Daly, Dr. John Hickman, Hugh Seppa, Bill Wilson, Greg Creegan, Evan Kumpala, John Hope, Bill Barnes, Roger Shannon, Jon Englund, Tom Edison, Lewis Wright, Roy Seeborg and Bill Cunningham.

ASTORIA
COLUMBIA RIVER BRIDGE
ASTORIA CLOWNS
LET'S BUILD THE BRIDGE
COME

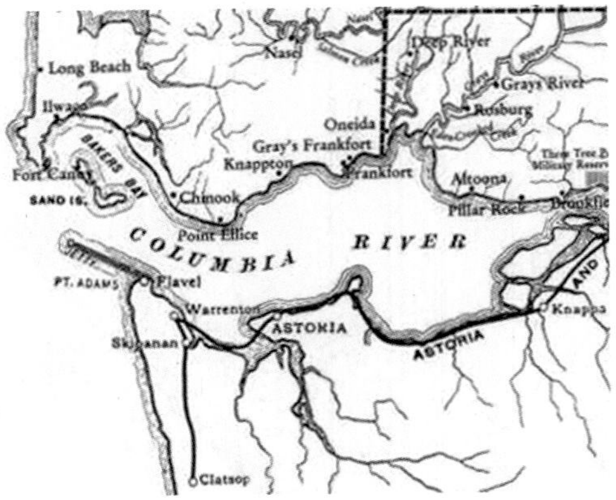

Ilwaco
Long Beach
Ocean Park
North Head

1966

North Head
Lighthouse
1898

In 1880, Henry Harrison Tinker purchased land in the Long Beach area intending to develop a resort. Tinker soon attracted travelers to the area by filling swamp areas, naming streets, by developing camping plots, cottages, and a hotel. Tinker advertised in Portland, Oregon newspapers, which drew more visitors to the coast. The resort was commonly referred to as "Tinkerville" until a post office was established in the area. In 1887 the area became officially recognized as Long Beach. In the 1890's a railroad and station was built in a small subdivision at the north end of Long Beach.

Emil Bjork is second from right, logging in Deep River. He later opened a dairy in Astoria and Seaside and brought the first Pure Pak folding milk carton machine West of the Mississippi to the area.

They were all old then,
The fishermen, the loggers, the long shore gang.
Old. They were Jens and George and Bill. And there was Enio, Vieko, Karl.
They lived in a cabin in the country. They logged and they fished.
They had a farm, big or small, with cattle and hogs and fruit trees and vegetables.
Water from the creek, light from the lantern, heat from the wood stove.
Not much water in the summer, little oil for the lantern, cut a tree for the stove.
Just straw for ticking in the mattress. Perhaps some glass in the window.
Rockin' chair on the porch but no time for sittin'
These are the folks that made this county. Workers. Hard workers.
They gill netted, gypo logged, worked with danger every day. Often staying in the logging camps, sleeping on the fish boats. They raised us.
They were the makers of Clatsop County. We know their names. We were raised here.
They were Finns, Swedes, Norwegians. They were Irish, Scottish, English.
They were Native American, they were European, they were Asian.
They were all old then.
Those citizens of Clatsop County...

by Michael T. Manion

Under all is the Land

B & W Timber Co.

Deep River

Our big adventure was to go on the ferry from Astoria to the Washington side of the Columbia River to Deep River. My grandmother knew all the Porters and the dock people by name, who would yell "Hey, Fannie!" You had to wait your turn for the man in the uniform to wave your car onto the ferry and they packed you in like sardines. As soon as the whistle blew and you were launched, you could get out of the car and go everywhere on the slowly floating ship. When you hit the pilings on the other side, you knew you were there. Something was lost when the ferries went away after the bridge was built connecting the two sides of the Columbia. People went alone after that.

A railroad was built to bypass rapids in portions of the Columbia River and down the Pacific Coast. The Oregon-Washington Railroad and Navigation Company, a Union Pacific predecessor, operated steamboats such as the Columbia and the T. J. Potter.

"AS ALL WE DREAMERS KNOW:
IT'S NOT THE WINNING -
IT'S THE GAME -
WE ALL STAY THE SAME -
IT'S JUST THE DREAMS,
THEY GET OLDER."

BY ROBERT JAMES WALLER
"THE BRIDGES OF MADISON COUNTY"

Kinsey Photo

Clark & Wilson
Logging Camp

Kinsey Photo, 1927

Grace & Bess McNeal were logging camp cook house children. Bess becme a cook herself, and is in the picture at left, top of stairs.

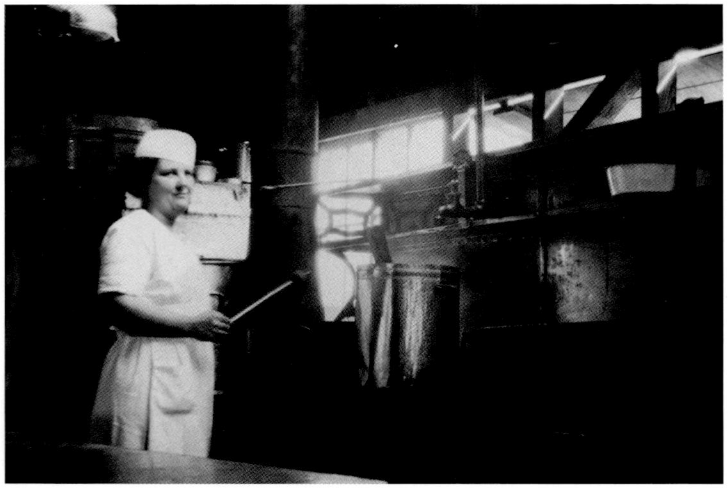

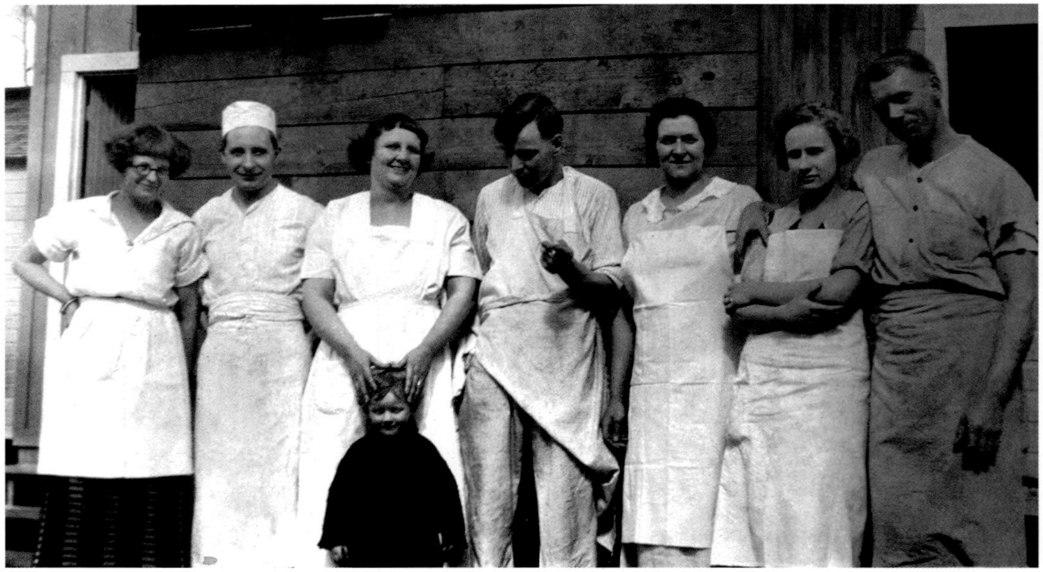

Cook Houses

Logging Camp Dining Room

Deep River annual carnival in Deep River, in the 1920's.

Crown Willamette Paper Company

The B-W Timber Company, a big railroad operation, set up camp on the old Clay Dooley ranch on the Nehalem River between Jewell and Elsie. Bill Hollenbeck was supervisor of a camp that housed some 500 men, equipped with living quarters, machine shop, cookhouse, family houses, office quarters and bunkhouses. A trestle had to be built to span the Nehalem River so eight miles of railroad could be built. Tidewater Timber Company out of Olney took over B-W properties, extended its mainline logging railroad from its summit camp to Jewell and beyond to the B-W holdings which it renamed Tideport. The Buster Creek line was abandoned and Tideport logs were hauled to the Tidewater dump at Olney. Hollenbeck struck out on his own, and his first crack at being a gyppo logger was on 160 acres of watershed timber owned by the City of Gearhart. Steam donkeys and trucks were used. During this time with the aid of a Seaside blacksmith, he devised a two-axle log trailer. The old single-axle trailers used at the time were limited by law to 17,000 pounds. Some of the old growth timber they were cutting had to be bucked into eight-foot lengths to meet weight limitations. The two-axle multiple wheel trailer revolutionized logging and speeded the demise of the logging railroad and gave birth to a brand new concept, trucking logs from the woods to log dump.

BY SAM CHURCHILL,
"BIG SAM"

Sophie and Emil Bjork in Deep River,

George and Bessie Hill, her sister Tillie Hendrickson (McNeal-Yates), and daughter, Bess.

The General Washington approaching the Deep River Landing.

Deep River was built on pilings because of the Columbia River tides, which came in twice a day. The sidewalks and roads were constructed of planks that were elevated at least six feet above high water. In 1917 Deep River had two stores, the Shamrock Hotel, a coffee shop, a community hall, a school, Pentti's Westend Pool Hall, daily boat service to Astoria, a logging railroad, and the Deep River Holy Trinity Evangelical Lutheran Church. There was a community sauna located between the hotel and pool hall. Many Finnish immigrants settled in Deep River and the surrounding areas of Washington, where there were striking similarities to life in Finland. Besides logging, Deep River had dances, a picture show, carnivals, picnics, and a girl's baseball team. They formed their own community of immigrants, isolated except for transportation by boat or ferry on the Columbia to Astoria and an eventual railroad.

Wakiakum Pacific and Clatsop Counties

Appelo's in Deep River was owned and managed by Mr. and Mrs. C. Arthur Appelo who were immigrants from Finland. It was the General store, U.S. Post Office, Telegraph office, and a gas station.

Emil Bjork, Deep River, fourth from left.

Kaisa
From Finland to Astoria and Deep River

Kaisa Eskola
Astoria, 1884

Kaisa, also known as Katherine, Eskola of Saloisten Pitaja Oulu, Finland, came to Astoria in 1883. She had worked for the U. S. Ambassador to Russia, William H. Hunt, who died in St. Petersburg while they were there, and the family had planned to bring her with them to America. She came on her own to join her brother, who had taken the American name of John Wilson, and her sister, Mary Matilda Broanda Johnson. She married Lars Frederick Hendrickson in 1884 in Astoria and they had four children. Lars Hendrickson purchased the Deep River property from the Great Northern Railroad; the deed was signed by President Grover Cleveland. When he died at age 36, she remained on the 600-acre property shown on the next page. In 1894 she married Matt Puskala, of Finland, and they had six children. She raised her family in Deep River, Washington. She traveled on the Great Northern Steamship from Astoria to San Francisco, sold door-to-door Watkins products, and lived in Deep River for 52 years. She was a charter member of the Finnish Brotherhood in Astoria.

Appelo's Post Office,Deep River, 1925.

Hendrickson-Puskala home, Deep River, 1936

Bridge with a View over the Columbia River

Oregon
and
Washington
connect
in 1966

John F. Kennedy In Astoria

In Astoria, John F. Kennedy and Dena Bondietti Marsch, with her Brownie camera and her teacher. Dena is the daughter of Sheriff Carl Bondietti and his wife, Joyce.

JFK found me working in JC Penney's during his campaign sweep through town. It was an interesting experience for a working student, clerking, small-town boy to have this storied senator, presidential hopeful, walk up to me, accompanied by a tall, fit, tanned and slender Marine bird colonel. He seemed interested in me, while not wanting to interrupt my work. Also straight up tall and tanned, he spoke a few words with me, shook my hand, and looked me over with some interest. This happened with him having walked to the back of the building where he found me fitting a pair of shoes. Just as quickly, he moved out to the front; it seemed he had come to see only me. I noticed no stopping as he made his way in and then out. Maybe it was my imagination, more likely it was because everyone else was out of view!

by John Marsch, Astoria

John Marsch's grandparents
and his mother in Finland

John Marsch, in the late 1940's,
Astoria Airport, and in 1963.

Marsch family in Astoria.

Tongue Point Job Corps was originally a U.S. Naval facility and declared surplus in 1961. The job corps for youth in the country was a dream of President Kennedy, and was established in 1965.

Celiast Coboway

Celiast

The Chinooks were on the lower Columbia River and the Clatsops were a smaller group of several hundred Indians who lived across the Columbia, on the Northwest tip of what would become the state of Oregon. The whites arrived in the 1790's and Chief Comcomly rose in power by gaining favorable trade relations with whites.

The Smiths settled in the isolated wilds of Clatsop Plains, which was Celiast's home, after she left a first abusive marriage to another white man in Fort Vancouver, and it was noted that she had always desired to come back to Clatsop which was to her, a sort of paradise.

The Clatsops remembered Celiast as Coboway's daughter, a distinction to which she could add, evangelist assistant and settler's wife. She was also known as Helene. Her son, Silas, said the Clatsops gave his family a royal reception in 1840, "As we neared the shore at the Clatsop village…the natives came running down and rushed into the water waist deep on either side of the canoe, and taking her by the thwarts carried her with her load".

They developed one of the areas most valuable farms and Solomon brought the first substantial numbers of horses and cattle to the area. He helped Oregon form a provisional government which sought admission to the U.S. and was Clatsop County's first representative to the State Agricultural Society and served in a variety of local offices before becoming a state senator.

Celiast was regarded by the Indians as having extraordinary "skookum tumtum", that is a bold heart. She is remembered in Oregon as the pioneer and Indian princess who under difficult and narrow circumstances made a place for herself and her children. Her most precious resource was bequeathed her by a people who barely survived her.

by David Peterson, "Celiast Smith: Pioneer Woman, Indian Woman"

Silas Smith was Celiast and Solomon Smith's son and he became a lawyer.

Solomon Howard SMITH (1809-1876) married in 1840, Celiast (Helene) Coboway was the first teacher in Oregon. He lived in Clatsop Plains near the Columbia, and was involved in politics on a state level.

Solomon Smith was a New Englander of good education and an adventurous nature. As a young man he had attended military school and had also obtained some medical training. He joined Wyeth's Expedition in 1832. Smith arrived at Fort Vancouver in October of that year and went to work as a teacher at the fort. While there he became involved with Celiast, the Indian wife of the Fort's baker. She left her husband and joined Smith at Gervais where he taught school for a couple of years. In 1836 he helped Ewing Young build a grist mill. In 1840 he formally married Celiast and they moved to Clatsop Plains near the mouth of the Columbia. He farmed, kept a store and was serving as a state legislator when he died. Solomon Smith was a mill owner, a state senator, and Oregon's first school teacher. His wife, Celiast, was the middle daughter of Coboway, who was Chief of the Clatsop tribe during the time of Lewis & Clark. They had a 640-acre homestead in Clatsop Plains.

Warrenton High School 1925

Bill Wallingford, second from left, was a Mayor of Warrenton and a grandson of Celiast and Solomon Smith. Incorporated in 1889, Warrenton was originally known as Skipanon or Upper Landing.

WARRENTON

Clatsop Plains Church and Pioneer Cemetery

Names of people and families buried In the Pioneer Cemetery are: Allen, Alymore, Basch, Bell, Bird, Bitterling, Bump, Bradbury, Brallier, Byrd, Callender, Carnahan, Chadwick, Chitwood, Cloutrie, Coffman, Coleman, Condit, Darland, Davidson, Day, Eaton, Eberman, Edens, Frederickson, Gaines, Gearhart, Geroy, Gibson, Goodwin, Gramms, Green, Grimes, Gunn, Hamblin, Hansel, Hess, Hill, Hobson, Holt, Houghton, Hume, Jeffers, Kimball, Kindred, Kingsley, Kinney, Knoblock, Laighton, Leinenweber, Long, Lonius, Loomis, MacEwan, Martin, Matier, McGuire, McKean, Morrison, Neal, Oeffler, Owens, Parker, Pemberton, Peterman, Poole, Powell, Powers, Prael, Price, Raymond, Robb, Roberts, Russell, Sackett, Sayre, Schrader, Shane, Shepard, Sherman, Shields, Shipley, Smith, Spellmeyer, Stafford, Stanley, Stevens, Stoddard, Sudheimer, Summers, Tagg, Taylor, Tee, Thomas, Thrall, Wallingford, Ward, Warren, Wheatley, Wingate, Wirt

Clatsop Plains Presbyterian Church was founded by Rev. Lewis Thompson in 1846 in the home of W. H. Gray, who was a member of the Whitman Spalding party. It is the oldest continuous Presbyterian Church west of the Rocky Mountains. The flowering cross in front of the church blooms every spring. The cemetery is the home of the first settlers in Clatsop County.

Josiah West family

Cousins, Russ Earl and Jon Johnson, with their grandmothers, Rose West, left, and Grandmother Webber, right. 1944.

Beerman Creek Farm, North Plains.

Beerman Creek Dairy

Josiah West and Beerman Creek Dairy

by Russ Earl

(Ed.Note: There were two Beerman Creek Farms, one at North Plains which is now the Surf Pines area, and one south of Seaside near Peterson Point, across from the present day Circle Creek Camp Grounds, which was all parts of this family property.)

The West family, beginning with Josiah and Lamira, was quite a large and interesting family. There were about 10 children. Many were women, and the name wasn't continued, but their influence was. Maude Prescott, Violet See, Rose West Johnson, Cecile Sauer, et al. were all part of the West family. Josiah West was a good businessman of the times and he had a creamery and was a serious landholder. The West estate took in, with variations, from the Delray Beach entrance north, somewhere in the Reed ranch world, and east including the cranberry bogs.

They were some of the early settlers. The West's were on my mom's side. The Johnson side had Rose, who was Jon's and my Grandmother, and they had early history at the south end of Seaside toward the Black Bridge area by Peterson Point. My Great-grandfather Johnson was the first owner. Sef Johnson's part was the Northern part of the Beerman area, the Peterson Point area.

Grandma Rose and each of her living siblings had a share of the 3,500 acre Clatsop Plains estate at Surf Pines when Josiah and Elmira West died, one being my grandmother and her share being West Hills Dairy, another was Paul See's mom, Violet, and on down the line. Sef Johnson's part was the Northern part of the Beerman area in Seaside. My mom's maiden name was Johnson and she was raised at the Beerman Creek farm because Rose West, from Clatsop Plains married Sef Johnson, whose heritage was south of Seaside. After raising their family, Rose and Sef sold the Beerman Creek farm to Bud and Elinor's Johnson and they moved to the Clatsop Plains area in the early 40s and built the south Surf Pines house and barn shown in the pictures. My parents bought the property from Rose and Sef after dad came home from WWII. Behind the windows in the barn picture was a "modern" milking parlor of the times, where the cows stood above so the milking section is at chest height with vacuum operated milking machines.

The 3500 acres went with deviation North of Del Ray to the Reed ranch, West to the ocean and East to and including the cranberry bogs. It is still Beerman creek. The name Beerman was the original Donation land claim name, the name of the creek through the place and the name of the Bud and Elinors dairy farm. The pictures are at Clatsop plains at the south Surf Pines entrance. The barn and house were built after the Beerman dairy. Sef and Rose moved to the plains and built the house and barn after selling to Bud and Elinor. The dairy at the Surf Pines house was named West Hills Farm, named by Sef and Rose, and the name stayed when my mom and dad, Ray and Pauline Earl bought the place and they owned and operated the dairy from 1945 to 1960. Now only about 30 acres remain.

The old barn was the Herman Sontag place. I used to make the hay at the old place when the Sontags were living there and she used to talk about the history. The old barn was there until Joe Camberg bought it, divided it up, and houses came into existence. Ed and Nancy Bates owned and operated Ocean Home Farm across from and south of the Delray beach entrance.

The house and barn, near Peterson Point is where Howard E. Johnson, and sisters, Nellie, Connie, and Dorothy were raised. Their parents were Albert and Annie (Peterson) Johnson. There were two brothers, Sef and Albert. The whole land tract in that area was one and the two brothers decided to divide it in half, Sef taking the north part and Albert the south where the white house and barn now sit.

Pauline Earl worked in the Shipyards during WWII when her husband was in the service.

Russ, Tracie and Steve Earl

Steve, Russ Earl, Richie Schroeder, Rod Inman, and Jim Lucas.

Rose West Johnson

Josiah West Home in North Plains

ALL ROADS LEAD TO THE SEA

In the late 1700's there were four independent groups of Indians, the Wahkiakums of Wahkiakum Co. Washington, the Cathlamets, along the South Bank of the Columbia, the Clatsops who lived along the south bank and Youngs Bay, and the Shoalwater Chinook, on the North Bank as far as Willapa Bay.

The Clatsop Indians

By Stanley R. Church and Jim Dennon

The Clatsops moved between winter and summer villages, Neacoxie at the mouth of Neacoxie Creek was the Clatsops inter-village where they moved at the start of the fall salmon run into the Necanicum and Neacoxie streams. In the spring at the start of the Columbia River salmon run, they would move to Neahkeluk at Point Adams, and Neahkstout near the present Hammond. The village in south Seaside, Ne-cotat and southward was inhabited by Tillamook Indians, thus the names of Tillamook Head, Clatsop Plains and Clatsop Spit correctly represent the former territories of each tribe. Clatsop Indians used the Neacoxie Creek as their canoe highway from Seaside most of the distance enroute to the Columbia River. At that time Neacoxie Creek was the outlet from Cullaby Lake, leaving the lake at the northwest end, flowed west, then north in a large loop on the oean side of what is now Smith Lake, then west about 50 feet, then south and flowed through Neacoxie Lake at Sunset Beach and on down to what is now Gearhart where it emptied into the Neawanna close to the ocean.

During the early 1840's a flood of immigrants poured into Oregon, spreading not only south and west of Oregon City, but to Clatsop Plains where the Indians were known to be peaceful. Nearly all of the Clatsop tribe territory was occupied by white settlers in 1851.

Jennie Meschelle

Philly Katata

Tansy Point Treaty

A treaty was entered into with the Clatsops which described their land as Point Adams, south along the Pacific to the mouth of a stream known to Indians as Yocklespahto, east upstream to its source, thence east to the summit of the Coast Range thence north to Saddle Mountain, thence northwest to the headwaters of the Lewis and Clark River, thence down said river to Youngs Bay; thence westerly and the southerly shore of the Columbia river to the place of beginning:

"Land being all that is claimed by the tribe."

The Clatsops were granted a reservation that was marked with a stake between Tansy Point and Point Adams, the stake being the easterly boundary of an old Indian burying ground. The Clatsops were reserved the right to fishing grounds at the mouth of Neacoxie Creek, and they could pass along the beach to and from their fishing grounds and Point Adams and were allowed to pick up whales washed on the beach. The treaty was never ratifed nor honored by the government. The last full blooded Clatsops, Kate Juuhrs, daughter of Chief Tostum (Chief Coboway's nephew) and Bob Salikee, died in 1926.

The pier into the Pacific Ocean, in front of the Hotel Moore, Oates Natatorium on the right.

Leo Marlantes, Anna Marlantes and Jenny Elias with their grandfather Louis Mehelis. Leo Marlantes became the principal of Seaside High School. Anna married Delbert Putnam and Jenny moved to California. All three graduated from Seaside High School and were cousins to Anna Mehelis McNeil.

"This is the grandfather who came to this country in 1903 and is responsible for all of us who are here. He worked in Chicago for awhile before moving to Seaside. My grandmother didn't arrive until 10 years later. Previous to that, my Aunt Tasia came with my Dad who was 12 at the time. Louis and Anna Mehelis had two daughters, Tasia and Sophia, and my Dad, James, who was the youngest. When my Dad was 24 he traveled to Greece, met and married my mother, Efstathia. I was born in Astoria because there was no hospital in Seaside, a year after he returned with my Mom. During the depression and WWII, my Dad owned and operated a tavern called the Gem Tavern which was adjacent to the Little Gem Cafe. Connecting the two establishments was an open doorway between the two. It is now all part of Dooger's. Leo and Anna Mehelis had 9 grandchildren and 23 great grandchildren."

by Anna Mehelis McNeil

Eleanor & Frank Kerr east of Broadway Bridge in 1927.

Anna Mehelis, Miss Seaside, 1950
with family friend, Fannie Hollenbeck.

Seaside

At the new Tides, Eleanor and Frank Kerr with relatives, Lyndel and Rose, 1927

Bobby, Joy, and Wayne Olafson

Sam Lee

Sam Lee, resident of Seaside, owned a laundry and was a professional clam digger. He would show people how to dig clams correctly on the beach on Sundays in front of the Hotel Seaside.

Lou Ann Kerr, Liz and Lorraine Allen in 1935.

A men's group during the 1920's put on a show.

An unofficial pageant before the Miss Oregon Pageant.

June Erickson

Majorette, June Erickson, leads the All Girls Band for Seaside High School in 1936.

SEASIDE
HIGH SC
BAN
TRAILS END
M.O.C.
Post
2934
SEASIDE, ORE

Bathing Beauties on the Beach in Seaside in 1936: Policeman, Harry Kemmerer,
June Close, Thelma Littrell, Jean Hickerson, Frances Haag, Margorie White, June Erickson.

June Close, Thelma Littrell, Jean Hickerson, June Erickson, Frances Haag, and Margorie White pose on the Seaside beach, 1936.

This picure shows Tom 'n Larry's when it was a restaurant before it became a candy shop. It was at the turnaround in front of the Montag Film Shop. On top of the Natatorium was the American Legion Club which later became the Pypo Club.

Leeown McMahon
in her Sweet Shop at
the Strand Theater.

Seaside Court
Joy Olafson on
front right, 1938.

Elmer Erickson
on the Prom after
high waves in 1941.

Broadway Cleaners
are Dave Scoggin,
June Erickson Cameron,
Miffy and Happy Erickson.

Elmer and Klara
Erickson

Seaside Bowling Lanes
on Broadway, early 1940's

G and A Logging Company partners,
George Gray and Ernie Alberti, 1955.

Ray Cameron, Wayne Smith,
Tink Murren, Pat Cook,
Marilyn and Bob Staley, in
Piggly Wiggly, 1948.

Gary Heller,
Penny Arcade
photo booth.

Vernon Olafson
cutting candy in
Tom 'n Larry's
Restaurant

The Penny Arcade
lunch counter with
Vern Raw and
Clysta Olafson
at the counter.

Ed Scott, who managed the Penny Arcade in the Summer of 1946, is shown in the photo jail with his daughters, Terri and Nancy.

Broadway Arcade

2007

THE PENNY ARCADE AND A BUS RIDE BY NANCY SCOTT

Though we lived in Seaside only 4 years, from the time I was 4-8, I have wonderful memories of our time there. Our neighbors were so patient and kind to Terri and me. Mrs. Jeffries, the beautician who lived next door taught us to play Canasta. I remember her inviting me to sub at age seven when one of her lady friends couldn't play. She also let us squeeze the orange food coloring throughout the package of oleo margarine when we made Toll House cookies together. Terri and I adored her. I remember Dr. Russell Parcher letting me take my favorite stuffed toy, Bozo the Clown, into the operating room when I had my tonsils out at age 7. He put Bozo to sleep with anesthesia first. Imagine such a thing today! My favorite memory is of my 8th birthday. I invited 20 friends to be my guests at the Penny Arcade on May 9th, 1949. We threw darts at balloons, knocked down milk bottles, rolled in a big barrel and had our photos taken in that jail. It was great fun. Sue, who lived a few houses down Downing from us, and I roamed at will, visiting the old salt cairn, riding our bikes to the Turnaround, and walking along the railing of the boardwalk to check the time on the sundial. We marvelled at the freedom we had. Though I'm sure we were forbidden to go into the ocean alone, I don't remember ever being afraid of anyone or anything during those years, despite the fact that my parents both had some serious health problems at that time. There was no kindergarten when I started school, and Terri was distraught when I was gone. Mom was really busy with our 3 month old twin sisters so, when I started first grade she made a deal with the city bus driver, who also drove the school bus route. He picked Terri up along with me every morning and, for a nickel a day, he let her ride his hour long route once around the city and then dropped her back at our house, thus adding some excitement to her day and giving mom a bit of a break. Seaside was a great place to be a little girl.

New Moms and their kids are welcomed to Seaside. Some in the picture are: Agnes Ward and Needham, June Daggatt and George, Dorothy Parcher and Gary, and Nancy and Terri Scott.

Lisa Smith's birthday, 1949

Ray and June

Margie and Donna Cameron

Donna Cameron and her cousin, Judy, are pulled by cousin, Dick Hunt, and friend, 1947.

Klara Erickson and Charles Bany lead a parade at the Turnaround.

Jerry Alto, Sharon Sapp, Judy Nutter, Tibby Utter,
1957 Seaside Bowling Lanes

Vern Raw, Miffie and Frank Erickson, Ray and June Cameron,
Happy and Marge Ross, unidentified, and Margaret Raw.

Seaside Train Station

Ernie Alberti and Ron Johnson

By the Prom ramp to the beach are Jim Delzel, Joe Fallon, Wendall Eads, and Bob Furrer.

"I will never forget when one of the contestants for Miss Oregon was asked what she would say about Oregon in Atlantic City if she was selected. She replied that she really wasn't from Oregon, that she was from Washington! She didn't become Miss Oregon."

Anna Mehelis McNeil, 1950 Miss Seaside

Gloria Irene Welch sits at he center of the Turnaround showing a map of the "End of the Lewis & Clark Trail" in the 1950's.

Frank and Eleanor Kerr drive a Miss Oregon contestant.

Miss Oregon Pageant Seaside

JoAnn Amorde was the first official Miss Oregon, crowned in Seaside in 1947.

On September 22, 2001, the life of 21-year-old Katie Harman changed forever. As she walked the runway to the pulsing applause of nearly 60,000 people and the wafting of the traditional melody, "There she is...Miss America," she thought of the women that had walked the same steps before her and the new steps that she was about to pave...especially in light of the recent September 11th attacks on America. Just like most little girls in America, Katie once dreamed what it would be like to be Miss America.

Miss America, September 2001
Katie Harman

Katie Harman
Miss Oregon and
Miss America, 2001

Margie Cameron's
birthday in 1949;
Some identified:
Donna Cameron,
Barbara Flukinger,
Margie Wilson, Shirley
Duncan, Kay Auld,
Twila Larkins, Sally
Barker, and Pat Carlson.

Bob Benson
6th Avenue 1950

Mrs. Van Gundy's Central School class
Some identified are: Janice Wheatley, Martha Beebe, Diann Smith, Rita Sabey, Sissy Savariau, Peter Wascher, Ed Sigurdson, Leslie Morris, David Anderson, Mike Rippet and Jon Johnson.

1950 Dahlia Court

Donna Larson, Queen and Jimmy Lyle, Prime Minister. Members of the court were Tommy Paterson, Susan Lowe, Danny Goozee, Karolyn Walters, Robert Uzelac, Doug Combs, Kay Auld, Rick Soller, Rita Sabey, Earl Dunn, Judy Vikturek, Barbara Jean Wyss, Corinne Holmes, Lorian Marlantes, Doug Loutzenhiser and Donna Cameron.

Seaside Girl Scouts

1949 Dahlia Court: Bob Uzelac, Margie Cameron, Sue Nimmo, Jim Roehm, Margo Smith, Kathy Perrin are some who are identified.

Bob Uzelac and Donna Cameron

HOTEL
SKOOTER
MIXERS

Railroad Tracks

by Karl Marlantes

"The railroad tracks used to run right through where Safeway, Rite Aid, and Pizza Hut now stand. The tracks paralleled Roosevelt Drive, a barely two-lane road with asphalt that had no shoulders. It is now a congested stretch of Highway 101 which used to run through the center of town past the Piggly Wiggly grocery store and Roth's Drug Store on the corner of Broadway. There were two dimensions to "the tracks" as we used to call them. They were a dividing line that divided east from west and they were an escape to adventure that stretched north and south. The tracks separated the Finn Town Gang to the east of them from the Avenue F Gang to their west. They formed the border across which one ventured, BB gun in hand, to ambush the enemy. Of course, the enemies all laid down their BB guns to go to school together at Central School, now destroyed and replaced by ranch-style houses.

The other dimension of the railroad tracks was north and south. I used to stand on them and look northward, smelling the creosote from the wood ties, wondering how it was that two parallel lines seemed to join, somewhere in the infinity of wonder that existed beyond the Neawanna railroad bridge where one could take blue gill, crabs, mud cats, and, oh the joy, the occasional steel head. The railroad tracks leading south went to a mysterious place called "The Y." It was there that the trains turned around, down by the shingle mill and mill pond. The Y loomed large in my imagination, because it was outside the boundaries set by our mothers. Those boundaries were south to Avenue H and Trodevick's Field, west to Highway 101, where Paddock's Texaco on the east side faced off against Mobil's Flying Red Horse across the road, east to the Neawanna River, where unknown to our mothers the secret raft was hauled up into the swamp grass, and north to Avenue A. "And don't you dare go downtown unless I send you." What loomed large for mothers seemed terribly confining to us – and the railroad tracks were the key to our freedom.

I first got used to the power and sheer fright of the big trains by putting bottle caps and pennies on the tracks, waiting breathlessly with my friends for the wheels to roll over them, then rushing from hiding to see what magical transformations had occurred, turning a three dimensional object into a two dimensional one – and the coins and bottle caps were always hot. Soon we were pressing things from marbles to nails to one aborted attempt at pressing a fist-sized beach rock that shot over our heads like a cannon ball.

From there it occurred to us that if we just got on one of the box cars, instead of just using them as stamping machines, we could lay flat on the roof and ride it all the way to the Y and back, and never – never set foot outside the boundary line. Yale Law school had nothing we hadn't figured out already. The first foray proved too successful. This was because as we rode clinging onto the small wooden cat walk that ran along the top of the car, feeling the freedom of the breeze, the smoke of the steam engine, we noticed that we passed something called Sunset Bottling right next to the tracks on Sundquist Road, now called Avenue S. Maybe we could get a free coke or orange soda? After all, it was us. We rode the car back to the switching tracks of the old brown and orange station and jumped off onto the heavy pebbles that supported the ties. (The railroad tracks were excellent for sling shot rocks and throwing stones in general.)

The next trip was a great success, we got orange pop – for free! We were also told not to come back. Fair enough. Since our magic carpet had long disappeared back to mom territory and the greater world beyond to the north, we walked back, doing that curious walk of the railroad tracks, having to either set your stride, just a bit too long, to land on the ties, or else hit the rocks between, which resulted in a general speed up and slow down augmented by minutes of balancing on the rails. All in all it was a great outing until my mother asked me how I got the orange all over my mouth.

Another great attraction of the railroad track was the water tower that stood next to it right at the end of Avenue E, which turned into a sand road heading south along the tracks to Avenue F and heading north until it turned into Roosevelt drive and black top at Broadway. When the steam engine came in, it filled its boilers there, and there was always a tremendous water spill that turned the ground around the tower into moats and waterways upon which popsicle stick destroyers and two-by-four battleships could be launched to be sent into a watery grave with the smooth stones so ready at hand.

There was a large oak tree on top of a sandy knoll right next to the tower. Next to that knoll was a brown and orange single-room shack that had a sheet metal chimney coming from the top that smelled of coal fire. Inside lived Newt. Newt smelled like coal and tobacco and always had at least three days growth of beard. He worked for the SP&S repairing track and doing other maintenance. He was also the king of the water tower and it was at his command that torrent of water issued forth that gave life to the engine – and floated our boats. Newt must have liked kids, because he would sit, smoking a cigarette, and watch us play. It was he that taught us how to double back the stalk of the small plant with a black bud on top, pull it back sharply, and whamm, the bud would fly at whatever you aimed it. I don't know the name of that plant to this day; that was because we had a name for it ourselves and never bothered to find another. We called it a Newt.

There was a practical side to the tracks. On cold mornings my mother would send my brother and me out with gunny sacks and we would walk along them picking up chunks of coal to burn in our kitchen stove.

There was also a romantic side. Spokane, Portland, and Seattle, presumably the cities serviced by the SP&S Railroad, seemed a little beyond the reach of even my imagination, but Astoria did not. This was because my Grandpa Axel and Grandma Aina lived there and it was no mystery at all. It was just too far away to walk to – not to mention beyond the boundaries set by the mothers. But the train! If the train could take us to the Sunset Bottling Company and the Y and the mill pond, just imagine what it could do if we hopped it going north. The problem was going south meant it turned around within a couple of miles and then stopped again where we got on. Going north was as a whole new deal.

I thought about it for some time and then came the blowup with my mother over something I can't remember. I informed her I didn't want to live with her and dad any more and I was going to live with Axel and Aina who treated me better. My mother, Elna, knew a lot about kids and gave us a lot of freedom – within the boundaries. She actually helped me pack. I put one change of clothes in a sweatshirt and tied the sleeves together. I made a peanut butter sandwich, which she helped me wrap in waxed paper. Then she said that Astoria was about eighteen miles up Highway 101 and that she'd call Grandma and Grandpa to expect me later that day. What she didn't know was that I wasn't walking. Heading north, I got up the iron rungs on the side not seen by the people at the station and lay flat. No one noticed. I headed for the train and waited until I saw the one heading north. I got up the iron rungs on the side not seen by the people at the station and lay flat. No one noticed. The train set off slowly, leaving the covered station platform with its large dollies with small iron wheels and waiting benches heading up Roosevelt Drive.

It was all quite exciting, until we started picking up speed. This was a whole new world to the run to the Y. The car seemed to sway like the top of the mainmast on the Cutty Sark. I buried my head in the sweatshirt and held on for dear life. I caught a glimpse of the Neawanna as the train rumbled across the railroad bridge and I could see the highway bridge to the left. Then we passed Mig's River House, that curiously round little restaurant that looked west over the Neawanna and out to the ocean. Now Mig's River House was out of town. I mean, it was the frontier. It disappeared behind me. I was now in full panic. It was the first time in my life that I understood about riding the tiger, although to my seven-year old brain I am sure that the metaphor was unformulated.

I was saved by the small secondary platform at Gearhart. Someone apparently wanted to go to Astoria – or beyond. I did not care. All I cared about was that the train was slowing to a stop. I scrambled off and stood there watching the train go to my grandparents without me. I sat down. About half an hour later, Newt came along. He called my mom, our telephone number, 586J, was drilled into my head, and she came and picked me up in the car. I think both of us were too scared to talk about it and we didn't.

Now, all of that is in my memory. The tracks are torn up and few of the people living in Seaside today even know that they once were there. When I see the Pizza Hut, the Safeway, the Rite-Aid, and all the other unique architecture of our times, and their parking lots, I feel sad. I want to see the tracks, the sand road beside them, the stretches of spruce and fir woods, thick with salal, and Newt, smoking a cigarette under the oak tree on the sandy knoll."

North of
Broadway
1940's

Joy Olafson on right, with a friend, Seaside High School, 1953.

The Sea Water Apartments above the Aquarium

Harmon Family, Peterson Point

Isabell Ross Harmon

Deborah Harmon Lee and family.

Central Grade School Seaside 1957

1st row: Gary Phelan, Pam York, Helena Rawls, John Perkins, Larry Smith, Betty Lassila, Roy Beam
2nd row: Kay Kemhus, Karen Erickson, Creddy Jones, Pat McCall, Lilly Schroeder, Bill Russelman
3rd row: Larry Spoon, Gary Utter, Terry Oakley, Harvey Lynch, Novella Crafard,
Ron Roden, Marlene McClure
4th row: Ginny Kari, Judy Seelye, Larry Schuchert, MaryLou Martin, Mrs. Helen Powers

Larry Seal, Riley Boyer, Loren Owen, Carl Larson, Tom Lenocker, Bert Cole, Bill Wascher, Bill Meador, Walt Van Campen, and Bill Inman. They were protesting the girls wearing the new sack dresses in 1958.

Creddy Jones as the Fairy Queen
in Medea at Seaside High, 1961.

Jim Auld, Norm Wilcox and Novella Crafard, 1961.

In February of 1996 I was lucky enough to accompany my husband on a business trip to Singapore. While he was working I took a day sightseeing trip to Malaysia. At a rubber tree plantation during a family style lunch the other tourists and I introduced ourselves and said where we were from. I was the only American in our international group. A Korean lady who was traveling with her husband and four year old son turned to me and asked where in Oregon I was from. She told me her favorite place in the world was Seaside, Oregon. Seaside, where the beautiful long beaches are. I responded that I had grown up in Seaside. I was so pleased to hear her tell of her visit to our home town, and the freedom her children enjoyed running along the ocean. She said how lucky we were to have all that beauty and space. Of course it was sunny and warm the entire week of her stay.

by Kathleen Kerwin Bean

The New Car

by Carl Larson

It was the winter of 1948 and I was eight. We lived South of the Bell Bouy and the house is about level with Highway 101 and the basement and garage is level with the riverbank. Every winter my parents would keep watch of the tides and weather report, so when the conditions were right, they would take the car out of the garage and park it up the hill near the highway because the river would flood and get into the basement. One day, somehow, they missed something. In the middle of the night my mom woke dad and told him that she thought she could hear water in the basement. Dad got up and looked and said yes, the water was in the basement but it was too late to do anything about it. He was so sick with the flu, he didn't really care at that point. Our new 1948 Studebaker was parked in the basement.

When it got light we looked and saw that the water had reached as high as the armrests. Dad called a wrecker. It had to park up by the highway to pull the car out. Dad had to wade into the basement to hook the tow hook to the rear bumper. They started pulling but it wouldn't come out. Dad investigated and found that the 4 x 4 fence post he used to stop the car from going too far in the garage, had floated up and jammed the front wheels so they couldn't turn. He had to wade to the front of the car in the dark garage in about a foot of water, and saw the post in pieces with a handsaw, so the wheels would turn.They finally pulled the car out and they did fix it. I don't remember what all had to be replaced. I do remember the upholstery had to be redone and they drove it for several years.

Fifty years ago, as a young bride,
She moved to Seaside, not knowing anyone,
Except her new husband.
Over 450 people attended her modern day Pioneers
memorial service,
as those who came knowing none,
as did the pioneers of 100 and more years ago.
Both came, did their job, made their mark,
and made Seaside a little better place.

by Mike Manion

Kay Manion

Judy Rapp, Kathie O'Dell, Ron Nakata, and Bill Weed are some who are identified in this 1959 photo.

Dick Rankin takes a break from his lifeguard job and playing the ukulele.

ON THE BEACH 1959

Janis Berg, under lifeguard tower with Don Campbell, Jon Johnson, and PK Hoffman.

Marty Chave, Donna Cameron, Trudi Schnitzer

Portland kids were part of the beach during summers in Seaside. Some of them had jobs as lifeguards or worked in the shops on Broadway. They were usually the ones who took the pictures.

"Seaside in the 50's and 60's was the perfect place to grow up. I look back now at how safe we were. I'm sure most of my generation can remember a summer time romance. I married mine in 1964. We get to Seaside about once a year, relive days gone by. One of the most important things about Seaside is the fact that a lot of us still have friends from grade school that we keep in contact with. That says something about where we are from and how life was growing up."

Audrey Powell Haynes

Kathie O'Dell, Lifeguard and Janis Berg

Kathi Kerwin,
Judy Rapp, Ron Nakata,
and Dick Rankin on left.

Next page: Joe Camberg,
Jon Johnson, Bill Weed, and
Kathi Kerwin.

Over the Pond to Seaside

by David Rix, England

"Growing up in a small seaside town near Newcastle, on the northeast coast of England, in the aftermath of WW II, the land across the Atlantic might as well have been on a different planet --- rich, totally inaccessible, and known to us only via the products of Hollywood. Six years of war had proved an enormous drain on our economy, and the everyday things we now take for granted, such as cars, TVs, and even refrigerators, were pretty well unknown to us then. My education, however, was first class, in a school where the students were selected by academic ability, and I was fortunate enough to have the opportunity to improve my command of French and of German, through student exchanges with their respective countries.

It was from this background almost half a century ago, that together with around 2,000 other teenagers, I sailed the Atlantic to live for a year in the "United States of 1960". I had the good fortune to have been given a placement with a truly great couple in Seaside, Oregon --- Earl and Audrey Weed, parents of the author of this book.

"Culture shock" is an overused term, but surely more than appropriate in this case. After our ship docked in New York, some of us flew on to Portland, where, with my new family, I stayed in a hotel, and ate in a restaurant. Three "firsts" for me. The next day, riding down Sunset Highway to Seaside, together with the author and her parents, I thought I must be dreaming, and would soon wake up.

To the new American Field Service Exchange Student, the Seaside of 1960 was a truly amazing place. Everything seemed "new" (scarcely surprising, as barely a century had passed since the arrival of Lewis and Clark). I was used to seeing buildings, still in use, anything up to 1,000 years after their having been built! And (aside from school buses) there was no need for the extensive and efficient system of public transport to which I was accustomed: everyone went everywhere by (enormous) car. In my 17 years I had NEVER before seen an American car.

David Rix

Within a few days of my arrival, I walked from Cannon Beach, back through Ecola Park and over Tillamook Head to Seaside. Beautiful. I was amazed that the people who lived there seemed to take their natural heritage for granted. And the more I saw of the Northwest coast, the more I came to love its wild and dramatic scenery. (Many years later, after having travelled widely in both the United States and Canada, I was able to appreciate even better my incredible good fortune, in being allocated to an area which I still consider to be "God's Own Country", the best place to live in the whole North American continent).

As time went by, it became more and more apparent to me, that despite often having the means to do so, people from around the Seaside area generally did not take their vacations in far-off exotic lands, being content to enjoy the delights on their own doorstep, and rarely venturing further than California or British Columbia. So it could be that perhaps my earlier observation was wrong: it was not that people take the Pacific Northwest for granted, but rather that they have an intuitive appreciation that nowhere else could be better, so why go there?

Almost 20 years have passed since my wife and I last visited Seaside. Writing this has made me realize that that is far too long, and it is high time we returned. You people of Clatsop County, pause for a moment's reflection, and realize just how lucky you are to live in such a wonderful land!"

The first lighthouse
in the Northwest,
was built 1856.

E. A. Weed, 1964

Crayton, Gordon and Les, Jr. in Venice Park.

Seaside Memories

by Les Morris, Jr.

Les and Orvie Morris with sons,
Crayton, Gordon, Les, Jr. and John.

To have grown up in Seaside, Oregon, during the forties and fifties is an experience I shall always treasure. That I was born in Astoria and then raised in Seaside is a direct result of WWII. Both of my parents were from Esmond, a small town in North Dakota of about six hundred residents, most of whom were immigrants from Norway. Sent by the United States Army to Fort Stevens in 1941, my dad along with my mom, who had not traveled far from Esmond, were overwhelmed by the many differences between Esmond and Astoria. Making the transition was a challenge, especially that of enduring the separation from dear family members back home. At the conclusion of WWII and with the arrival of some of their family members who received jobs at various businesses in Astoria, my parents decided that the area had much to offer and soon settled in the relatively uninhabited development south of Seaside: Venice Park, the initial location of many of my childhood memories.

My recollections of my youth are many and wonderful:

•Standing with mom and watching dad and his soldier colleagues parading on the marching field and saluting the American flag as it passed.

•Enjoying life in Venice Park where my parents build their fist home at 2501 Neawanna

•Going to the nearby beach nearly every day, weather permitting! to play, make sand castles, and catch crawdads.

•Walking to the Indian Place, a small grocery store on Holladay Drive in Venice Park, and also walking to town to shop at the two grocery stores on 12th Avenue, or at the old Safeway which faced Holladay Drive, or to the Piggly Wiggly store.

•Being thrilled when a huge circus camped at a vacant lot near the Indian Place and peeking under the tents to watch the preparations for the big shows.

•Being in awe of a group of gypsies who annually camped along the Neawanna River in Venice Park and enjoying their color, dancing, festivities while at the same time making sure no one was kidnapped.

•Having as friends and playmates my brothers, Gordon, Crayton, and later John and the Perrigos, especially Judy, Janet, and Jane, and Jack Mouncer.

•Getting my first used bike and going to the gas station on Holladay Drive to put air in the somewhat deflated tires, but only to be deflated myself for I put too much air in the tire and it popped.

•Having my first job delivering newspapers to homes in Venice Park.

•Experiencing an earthquake while playing on the vacant lot, now filled with homes, between our home and the Perrigos and all of us children thinking it was fun riding with the earth, but distraught when our parents quickly gathered us up and took us away.

•Attending school at Central, Broadway, and Seaside and receiving an education throughout the system from top-notch teachers. (Thank-you!)

•Attending Sunday School and Church at the original Our Savior Lutheran Church, a house-turned-church, and then for most of my childhood and teen years at Faith Lutheran Church, an experience that helped shape my values and influenced my choice of careers.

•Taking and enjoying our Sunday afternoon rides throughout Clatsop County and then delighting in a treat of ice cream and cookies at the end of the ride.

•Driving the "old" winding road to Cannon Beach to purchase meat from a store owned by someone from the church.

•Marveling every year at Christmas time over the beautiful outdoor decorations on the homes and in the yards.

•Cheering as Santa drove down Broadway to the Times Theater where all of the children were invited to a free Christmas movie.

•Receiving twenty-five cents, ten cents for admission and five cents for a treat, to attend a Saturday matinee at either the Strand Theater or Times Theater.

These are but some of the many and wonderful memories of an early chapter of growing up in Seaside where I was always safe, secure, and nurtured. With gratitude to my loving parents, Les and Orvie and my younger brothers Gordon, Crayton, and John, I can affirm that they, along with Seaside, provided a terrific setting for creating memories buried deep within my soul and ones I shall cherish forever.

Labor Day Riot of 1962 by Jerry Alto

It was Labor Day weekend in early September, 1962 and it was hot. Seaside rarely gets hot but there was no wind and the humidity was high. The air was muggy. The town was packed but not with teenyboppers. In those days college students, many who had worked all summer, liked coming to Seaside on Labor Day weekend to blow off steam before they went back to school. It was Party Time!

As I remember it by late afternoon there were just too many cars and people on Broadway (Seaside's iconic cruising street) so they blocked it off for car traffic. Everyone parked their cars and upper Broadway became a mass of milling people. There were a couple of taverns on upper Broadway (Trails End being one) and the beer was flowing. The partying was overflowing into the streets. The inevitable happened. A fight broke out in the middle of the street. At that time Seaside only had three or four full time policemen, the Donavan Brothers, being two. Two policemen tried to move in a break up the fight and the crowd held them back saying, "let-em fight." The police's only choice was to back off. Well, the stage was set. Thousands of college kids fueled by alcohol packed into a small area. But this wasn't a planned event so everyone was just milling around. Then a guy shimmied up a pole and shouted to the crowd, "Let's go get the life guard tower!"

A hundred or so of the crowd headed for the beach and toppled the life guard tower. I remember when it went over it landed on some guy and broke his leg. There was a bit of hesitation but they regrouped and carried the tower up the turnaround steps and proceeded down Broadway. By this time it was starting to get dark. The local police had mustered all the city employees and issued them riot sticks for crowd control. A fire truck was dispatched to try and control the rioters and they headed to the corner of the Times Theater and Penny Arcade. They started spraying down the crowd in hopes dispersing them. The rioters moved in and either slashed the hoses or they ripped them away, and began spraying the firemen. An Oregon Liquor Control vehicle was spotted and overturned by the crowd.

Meanwhile all the Broadway business's had closed shop. A group of rioters had ripped loose one of the public benches that were on the prom and they were using it as a ram to break down the door of the Trail's End Tavern, which had cases of beer stacked to the ceiling. A couple of city employees tried to step in and stop them but were just shoved out of the way. But next door to the tavern was a small restaurant owned by a feisty lady named Oney Camberg. Oney was an imposing figure with an even more imposing set of vocal cords. She got in the face of these guys with some hard core profanity. Stopped them cold in their tracks. They dropped the bench and got out of there. This Seaside Riot was the real thing. Nothing planned, just a combination of circumstances that came together on a hot Labor Day weekend in 1962. The following years punk kids tried to keep the riots going but they were nothing close to the real riot of '62.

Cottages and Castles

The Flower House

Toland's Lobster house

Plummer House in Warrenton now Antiques by Renee'.

The Gilbert House

This had been Mrs. Flemming's Kindergarten in the 1940's.

Captain's House

Montag house

Rogers Hotel

Seaside has always welcomed tourists. In the city telephone directory in 1950 there were 300 motels, rooms, cabins, hotel rooms and cottages for rent. The Rogers Hotel was on Broadway next to Ter Har's.

Berg house on South Downing.

Scoggins' house at the Tides.

A COTTAGE NEAR THE SEA

The Tides 1927

This house near the Aquarium has had pictures taken of it over the years by Kevin Kent whose parents both grew up in Seaside.

Par Tee

Clatsop Village

Ancient Villages of the Clatsop Indians excavated in the 1960's during a dig by the Smithsonian. The properties are referred to as the ParTee Site, as it was on the Seaside Golf Course Property and the ParTee Room was the restaurant on the golf course above the present clubhouse. Indian burial grounds were near Cartwright Park on the Necannicum River by Avenue U where the custom was to place the deceased in their canoe on stilts along with their belongings.

The World's Largest Sitka spruce tree came down at 750 years old in the storm of December 3, 2007.

The Cove in the 1950's before the road was paved.

Lowry Collins, Marilyn Staley, Nona Mooney, Adele Wood, Audrey Weed, Helen Cummins and Jean Ter Har, 1967.

Tsunami 1964

On March 27, 1964 we had a surprise Tsunami after the earthquake in Anchorage, Alaska. There was a club in downtown Seaside, the Pypo Club, for under 21 year olds. It was in the Turnaround building above the natatorium in the space where the Legion Club had been in the past, but it had moved to ground level near the Necannicum River. We were dancing to live music when at about 11:30 we felt water on our feet. Thinking it must have been a water main that burst, we went to the door, and there was water rushing on the street and sidewalk. Barbara McKnight had an open jeep parked by the door and we jumped in it and shouted into the club, "everyone to McKnight's place." They lived in Clatsop plains. Her mother had food and pop on the dining table right away, as if she knew we were coming. She couldn't possibly have known that around fifty kids would show up that night. We were used to going to high ground whenever we heard a tidal wave warning but there was no blast. We always thought it would come over the town from out of the ocean, but it didn't happen that way; the rivers overflowed. Without missing a beat, we knew just what to do, and we all spent the night with our ever ready hostess at McKnight's place in Clatsop Plains.

Seaside to Viet Nam and Back

by Crayton V. Morris

Seems strange for me to be writing concerning my enlisted/officer years and my tour in Viet Nam as a Helicopter pilot 41 years later. I enlisted in the Army March, 1964 and went to Ft. Ord, California. I didn't realize how difficult basic training would be, but I graduated then went to Ft. Rucker, Alabama and became a Flight Simulator Operator at Ft. Leavenworth, Kansas. I kept pilots up to date on how to fly via instruments. Then, in 1965 I decided to apply to flight school to become a Helicopter Pilot and with the help of a general I was on my way. Must say that going to flight school was absolute hell with all the harassment and spit shining. I graduated on June 19th, 1966 and our graduation speaker was General Westmoreland. He actually stopped me in the hand-shaking line and asked, "How old are you son?" My knees started to shake as I wasn't expecting that. I turned 21 that day, became a Warrant Officer and a Helicopter Pilot.

On leave before going to Viet Nam, I spent a few weeks with my family in Seaside, and in Fisher, Minnesota. I was assigned to the 15th Medical Battalion, 1st Air Cav Division at An Khe. We were located in the Central Highlands of South Viet Nam. My first mission was to go into an area and pick up some American soldiers who had been dead for awhile. My chopper also received a bullet hole in the door. It was then that I realized that I was involved with "one hell of a war." Our mission was to go into an area and pick up both the wounded and the dead. Our motto was "So that others may live."

After about 2 months I was transferred to C Troop, 1/9th Cav., 1st Air Cav Div. Our unit was nicknamed: "The Suicide Squadron" as we were always the first ones to go into a new area with a platoon of infantry men. Most of my time with this unit was spent living at LZ's or landing zones. I would build a "homemade tent" or living area (hootch) out of wooden ammo boxes. Dad had sent me a hammer and a saw! I flew "slick or lift ships" and would fly the 2nd squad into these areas. After dropping our men in we would fly to "secure" areas with our choppers waiting to go back and pick our infantry. Dad also made sure that I had the Oregon State flag. This was very special for me and I flew this flag every day outside my hootch! I believe it came from Gov. Hatfield or the American Legion. We would eat C rations warmed up with JP4 fuel and to get a nap in before flying back again. In some cases not all of our infantry would be coming back to our choppers alive!

Just before Christmas 1966, I spent a week on R & R (Rest and Recreation) in Tokyo. I enjoyed being there and away from the war. Tokyo is a city that to me was extremely similar to New York City with all the hustle and bustle going on and very cold. When I arrived back in Viet Nam I saw the USO Bob Hope show at An Khe, our base camp. I was scheduled to fly one early morning and could not find my flight helmet and at that time learned that my best friend flight school had borrowed it and was killed during a night mission....he was my best buddy.

I found out the Army was looking for pilots to go down to the Mekong Delta area so I checked that out and was infused to that area as "Assistant Helicopter Maintenance Officer" at Soc Trang. I thought this would be a piece of cake flying helicopters just coming out of being repaired or routine maintenance. But, I would test fly these helicopters at night and at night those damn red tracers (bullets) would be shot up at my helicopter and was a bit scary. Another part of our job was to go into an area to rig downed helicopters to be brought back for repairs. Our chopper usually would have only a helicopter gunship or two flying around to protect us. To be on the safe side, we would fly only 3 to 5 feet off the ground call contour flying and going about 80 knots per hour.

While at Soc Trang I was able to look up an old friend of mine from Seaside, Oregon, Herb Kendall. He was in the Navy and stationed in Saigon. This was very special for me being in a war and to be able to see and talk to an old friend from Seaside! We spent a couple of days getting re-acquainted and seeing the sights in Saigon.

In July, 1967 I was finally on my way home to the good old USA! We stopped in Hawaii for re-fueling and I could sense a real change. Continued on to Oakland and my brother, Les, Jr. picked me up at the military airport and on the way home he took me to the Haight-Ashberry area and that's when I realized that my country had really changed. I had only my military clothes to wear so instead of my brother's car being damaged because I was in it, I asked him to immediately leave this area. This is when I woke up to what had happened the year I was in Viet Nam with protests against the war, etc. Most of the returning military personnel kept our mouths shut/quiet as to what we did over there. It wasn't till the 90's that some young people where I was working in Portland asked about my experience in Viet Nam and to see some of my pictures. There are hundred of thousands of us that need to talk about the war and I know a lot of people are interested. I was lucky growing up in Seaside because the people at home were very cordial and happy that I was finally home.

2nd Lieutenant John Alto
Rifle Platoon leader
US Army

John Alto has published a video of photographs from slides he took while in Cu Chi, Vietnam in 1966-67 that is included in the new book by Carl W. Burns, "Centaurs in Vietnam - Untold Stories of the First Year." They belonged to the 3rd Squadron, 4th Cavalry, 25th Infantry.

First Lieutenant Karl Marlantes, USMC, receives the Navy Cross

Seaside had many who volunteered, were drafted, and served during the Viet Nam War. There were decorated heroes among them, and some are now telling their stories forty years later for the first time. There are too many to list; that would be another book, however, they have heroic and inspiring stories.

Among the 58,200 who gave their lives, Seaside lost six of their own. Names of those are inscribed at the Seaside Library:

John Bossom, Melvin Hebert, Jr., Richard Olson, Carl John Green, Jr., Rick Cavanaugh, and Gordon Zimmerle.

Helicoptor Pilot
Warrant Officer
Crayton Morris

Capt. Rick Olson, USMC

Gearhart

From the back of a postcard, July, 1923 -
"This is the new hotel. It does look swell."

The Gearhart Hotel and WWII
By Annalou B. Thomas
A Temporary Home at the Beach

"The Gearhart Hotel was then in its fourth decade and it was like an English country manor. Standing on the dunes of the Pacific, it was a magestic structure."

Ten Navy wives were making the trip south to be with their husbands in each port. Astoria did not have accommodations so the Captain went to the manager of the Gearhart Hotel to ask to rent rooms. Since it was late April, the hotel was open only for dinner - no guests for lodging before June. The manager finally agreed that if each wife were responsible for cleaning her own room and bath and helping with maintenance chores it would be feasible. The wives were allowed to go to the great old kitchen to prepare their own breakfasts and at noon we walked on the beach.

The ship's doctor was married to Claudette Colbert, the well-known movie star, who worked the switchboard as her extra duty. A highlight was Jack Benny, Mary Livingston and Rochester stopping one morning for coffee. The entertainers were driving to Seattle to entertain the troops. Gathered around the fireplace, the Benny's put on a rehearsal of the future show. This was such an unusual period in time and it is important that the Gearhart Hotel be remembered for the hospitable part it played in the history of World War II.

Gearhart Hotel Lobby

"A Frigate Like A Book"

by Sandy McCall

Once, in 1972, there was an April evening when four friends stood in front of some salvagers' scaffolding and argued whether five gallon gasoline drums in the Mattress Room or five gallon gasoline drums in the Vent Room would be the best way to burn down the Gearhart Hotel.

Other than a certain expertise in large beach fires, none of the four had any particular background in the incendiary arts. They weren't arsonists, but few arsonists ever had greater motivation. In several days tractors and a wrecking ball would arrive in Gearhart to tear open the sides of a great ship that had been, more than the beds they slept in, the center of their existence.

In the end the four friends didn't set the fire for a number of reasons. Fear of legal consequences wasn't high among them. One of the friends was facing a political arraignment in a Federal court; two others were brothers and very young husbands with babies. Yet in the end, the friends stood down because the fourth among them had a father who owned the place and they didn't want to cause him trouble. From a very young age they had realized that John Osburn was one of the best grown ups any one of them would ever know. Now, after years of struggle, John was at the endgame. Season after season he had made magic simply keeping the hotel alive. His trials had given employment to dozens of people and cheer to thousands. His gift to the four friends had been even more personal: A safe and ecstatic childhood that children in Dresden and Hiroshima could only gasp and pray for.

From October to April, the friends concurred, the hotel had always been theirs alone. From the time they had learned to walk they had learned to run wild through the empty hallways and great spaces of winter light and early darkness. Learned early that there was a secret key to the pop room where 7-Ups were stored, that the same room held four-wheeled furniture dollies that made such excellent bumper cars. In the old hand crank elevator they had discovered the mechanics of a switch that could throw off the lights and freeze the carriage between floors. That accomplished, the hand crank could be thrown left and right in a masquerade panic, a properly stunned voice could announce how the cable had once broken. But all that was ever broken were other children who would eventually be released on the fourth floor to walk back down to the lobby weeping horizontal tears.

The friends remembered learning other things that occurred in winter, things they didn't understand at the time, but reasoned out later; settings in hallways on December afternoons, the light dying as the great place shook under sou'wester storms and low voices were heard behind some of the ship's doors. Voices taking advantage of the emptiness, voices of big people who didn't want anyone knowing where they were. The hotel was a creature of the seasons as much as any 18th century British village. To the four friends it was bitter irony that the place was going to die in spring. That was the time it had always come alive, when people came back, like the daffodils.

The arrival of the first four day convention in April was like visiting a port of call after a long voyage. A boy would walk into the lobby and displayed in front of him was the bric-a-brac of the Oregon State Realtors or the Portland Car Dealers or the Practical Nurses of Eugene. Among the most welcome were The Associated Merchants of All Things Electrical who trafficked in stoves and refrigerators that came in large boxes that doubled as excellent execution chambers. Outside the hotel the boxes were loaded with some double dared nine-year old, then pushed gleefully off a steep hill toward the dunes in an end-over-end wailing of thumps and corrugated doom.

From April to October the conventions came and went in precision. The youth of Gearhart as yet had no interests in Realtors, Nurses, or Electricians' Brotherhoods, but the Magicians' Alliance was another story. Always a great hit, the magicians were something to skip school for in the vague hope of being sawed in half. To make up a truancy day a boy could always make attendance at the Oregon Educational Association whose display tables usually contained useful booklets with the test answers to Seaside High School's algebra and geometry courses.

As April turned to May and nine-year olds turned twelve, the hotel offered its pool as a forward observation post into the future. Older brothers and sisters— always mild up to now— were suddenly strutting their stuff. Two-piece bathing suits appeared on babysitters while former center fielders sizzled high off the diving board like over ripe marlin. The penultimate ritual of the swimming pool was something called "Horse," Gearhart's answer to the Aztec frenzies where once quiet girls climbed eagerly up on the wet shoulders of once quiet boys to take up combat in a touching, clutching, lubricious, knee-locked swirl of the senses. Eventually all bodies fell down, along with the fences.

If any parents noticed, no one ever seemed to care. For several generations Gearhart Beach in general, and three successive Gearhart Hotels in particular, were romantic stages worthy of the Milan Opera House. The ship had exactly one hundred rentable rooms on four floors. Every one had seen its share of scenes. Any family who ever came to the coast has at least a few members with safe deposit memories of life behind the doors.

Spring would turn to summer. Along with the conventions, the month of June saw the return of the Wounded Deer, the certifiable single guests who were naturally drawn to the beach for a freedom they could find nowhere else. Male antique dealers shared the same room because no one cared. The painter sometimes drank turpentine and water because no one watched. The lonely and famous writer of cookbooks who had found his first friends as a boy creating beach fire cuisine in the 1920's sometimes returned, not out of nostalgia, but of gratitude. Almost always the young man who carried their bags was the pompadoured bellhop with a temper problem. In his solitary role as winter caretaker it was he who spent his late afternoons lurking in alcoves or chasing the four friends down the fire escapes like a hound of hell. The bellhop's kennel was on the top floor and the friends knew with utter certainty that he would never escape it, if only for the simple reason he had murdered at least a dozen boys just like themselves. Their bodies were somewhere out on the dunes. It was just that nobody had found them yet.

That was only the A group. John Osburn's hotel ran the alphabet. And like his heart it was never a place that dwelled on differences. The ship had room for the bellhop and the boys, for everyone, actually. Even when it was fully occupied, there was always a folding bed for the used up millionaire who couldn't pay his bill, a cot for the lost jockey, the runaway teen, the newly released and hope-to-stay-there. The Gearhart Hotel ran on adventure, not money.

Every summer day presented new joys. Most welcome on July mornings were the occasionally married women who carried drama in their suitcases. Diamoned wives of welfare timbermen, conveniently absent; barefoot contessas who counted the days until the cowboys from the rodeo appeared in the Driftwood Bar. Wives of writers, wives of fishermen, young wives unsure of choices they had made and getting away for a few days to think it out. Two of the four friends had worked summers as night clerks at the front desk, noting the adult coming and going of the bar at two in the morning, putting on earphones and working the blinking lights and pull cords of the old fashioned telephone desk through the night. It was Closing Hour 101; plugging black plug into red plug, lives of a known world into lives of the unknown.

"But was there ever romance for you?" memory asks with advantages, seeking answers as far away as Agincourt. For the four friends it had always seemed so. With July the nagging rain was finally gone and sunny mornings meant nothing if not the newest pleasure of them all: of running into the lobby and stopping cold at the sight of age-appropriate girls from a place called Somewhere Else.

The lobby had its own ritual for Gearhart's saplings. New faces would be seen sitting expectantly on the lumpy couches near the fireplace. A shy smile begat a shy hello in response. "What high school?" begat the first game of ping pong on the Sun Porch. Ping pong begat swimming, which begat "Horse," and maybe going back inside the building for a more personal talk along side the windows in the Bikini Club. Walks on the beach begat driftwood fires which begat songs of the Kingston Trio. And, on occasion, the Darwinian imperative of adolescent desire begat something even more Darwinian: a turning of the corner into teenage honesty and something lifelong. Beach loyalty that never failed and could be counted on to the grave: a bond true as smashing surf.

Boys lie. Memory lies. And sentimental male memory lies most of all. Generations of beach houses have their fables of the hotel, of how summer friendships sometimes advanced in the darkness of the hotel's more opportunistic locales. There were stories of love in the elevator, of itching embraces on carpet piles stacked under the 16 foot sunglasses left from the War, or the wretchedness of Lord-of-the Flies voyeurs clamoring over acres of forgotten furniture to a peephole in the Vent Room. Dozens of stories. Hundreds untrue.

The most apocryphal fables of all concerned a hotel underbelly called the "Mattress Room," a forgotten non-descript place at the foot of the old stairways beneath the kitchen. It was, in fact, a paint starved, high-ceilinged storage area stacked to the toe-top with a mountain of oily denim mattresses.

At its summit there was perhaps three feet of horizontal breathing space. It was an act of faith for a nervous stranger to enter the room and climb to the top of that mattress mountain, especially with a boy she hardly knew. Even more faith was needed to stay there, if for no other reason than the room was adjacent to the hotel's great furnace and the heat through the walls could turn the love nest into a steam hole right out of Dante. Stories abounded that the temperature was so hot people had to take off their clothes just to endure the heat. More accurately, heat and peer pressure brought most couples off the mountain safe and mostly intact. Mostly intact were the operative words. Mostly intact was the way the big ship bought you home.

The four friends who didn't burn down the Gearhart Hotel went on to their destinies. Often, they see each other and always ask the question why were they so lucky. They repeat the fables and their minds walk through the hotel as if it were a blueprint before them. Sometimes they find themselves in the Sun Porch, looking though a picture window down to children running around the pool. They turn to another window. Beyond the horizon the kind old sun is setting itself down towards Japan.

In their mind's eye they walk to the other end of the room and there is the ping pong table of their childhood. At each end of the wooden table a V-shaped cut has been whacked out by hundreds of players and thousands of games. The players fouled up. They whacked the table. The V shaped cutouts grew through the years, testaments of competitiveness and impatience, the primacy of teenage ego trying to impress somebody.

Some people, having been born in the century of the World Wars, might be tempted to see the destruction of the ping pong table as a small metaphor of its time. An uncontrolled foreshadow of the Era of Selfish Lives. That would be a mistake. The V's are really emblems of how kindly John Osburn lived in his sea salt world. He endured this vandalism as he endured everything in his life, with patient understanding. Disfigured cuts in plywood are John Osburn's ticket into heaven. Like the arrows of Saint Sebastian they indicate how true his life really was. Four men will never know why he was sent to be their protector. But always through their lives they have prayed in thanks that he was.

Sam, Preston & Jamie Devereaux

Gearhart Stories

My family was one of the fortunate ones whose Grandparents, and Great Grandparents had houses in Gearhart. They were next door to one another on the south side of 1st and north Ocean. It is now 5th Street. Next door to the west is a house now owned by Randy and Jeff Miller, and in the late 1940's and 50's across the street on the north side of 1st was the beach house of Judge and Mrs. McCollough.

Now in addition to their beach house, the McCollough's lived in a very fine home in Portland's west hills. Their home was on Ravinsview Drive and their oversized lot went down the hill all the way to Gerald Avenue, where we lived directly across from their stone retaining wall. They had beautifully landscaped gardens with pathways winding down their terraced yard. Lots of full sized rhododendrons and azaleas, box wood hedges, cut flower beds, large fir trees and great hiding places for naughty children to seek. The McCollough's were probably in their 70's at the time with very little patience for rowdy kids in their yard.

In the spring, as soon as it quit raining all the time, mother would kick us kids out of the house and we would play with all the other neighborhood kids. Our house was kind of the meeting place because it was a very quiet street with very little traffic, if any, and we had a turn-around in front of our place with a manhole cover in the center. This served as "home" for games of hide & seek or kick the can. And, of course, the McCollough yard was the place to hide. After a week or two of pleading with us not to use their yard, they would pack up and come to Gearhart for some much needed rest and to get away from us kids.

The day after the Rose Festival, mother would load us four boys into the car for the summer and also head to Gearhart. We would pull into the drive of my Great Grandparents' house, and, after suffering over two hours of getting to the beach, we would immediately start racing around the neighborhood, which also included the McCollough house and yard. The Judge, anticipating our arrival, would already have their bags packed. As soon as we showed up, he would load the bags into the car, and no matter what time of the day it was, they would head back to Portland for the whole summer of peace and quiet, knowing they would not have to deal with us until after Labor Day.

by Preston Devereaux

Hotel Gearhart #2

BY TOM LENOCKER

After the big flood of '48 a big black rubberized tank showed up in the upper end of the little creek that runs north and south in Gearhart. We theorized that it must be some sort of storage tank, possibly fuel tank from an airplane. Whatever it was, it made a great raft, It could hold four or five adventurous souls. It was a fishing platform, a pirate ship, an exploration vessel, for the upper reaches of the creek, not to mention a way to reach the other side of the creek. It stayed around for a couple of summers and disappeared. We had all sorts of theories, all sinister, but most likely what happened was that a high tide in the winter floated it back to sea, where it was deposited somewhere for some adventurous boys to live out their fantasies.

One thing that the kids in Gearhart grade school could look forward to was the annual Dairy Ice Cream Convention at the old Gearhart Hotel. For about five or six years, late 40's early 50's, the dairy industry would get together, the big fall out was a benefit for the kids, because they would introduce the new spring flavors of ice cream, and the left overs would make it to the school. If some of us found out early, we could go the hotel and sample before the stop at the school,

GEARHART STABLES, 1950'S

BY BOB FURRER

My dad was a logger and worked around Seaside and Astoria in the early 1930's. In 1935, we moved from Young's River Road, about a mile south of Miles Crossing, to 956 5th Avenue in Seaside. I still recall the so-called 1936 riots, which were really the conflict between labor and management in the effort to unionize the Crown operation. In the fall of 1936, I began the first grade in the former Central School building - Dick and Jane and Spot! This was the time when the Cartwtrights ran the 12th Avenue Market, where my folks did most of their grocery shopping. I remember my first airplane ride, off of the hard sand beach at the end of 12th Avenue. It was in an enclosed cabin bi-plane, where my mother sat in the seat behind the pilot, along with her niece, Alma and held me on her lap. The price of the flight was a penny a pound for passengers. I also recall the weekend event of going to the train depot in the summer to wait and watch the Daddy Train pull in from Portland.

The Chili Bowl was operated by my Aunt Josie Crain and her partner Bertha Doaks, wife of Oliver who had Doaks Place Beer Parlor. In the late '30's, I would take the bus from Hillsboro to Seaside, and spend summer with my Aunt Josie and cousin Lois at their Oceanway residence and hawk News Telegram papers along Broadway. My favorite deposit of my hard earned money was the little candy-ice cream window in the Times Theater building that opened out on Broadway. Here you could buy a popsicle for a nickel and save ten wrappers and trade them in for a free Popsicle.

My first job out of college in May 1957 was as a Planning Technician for the City of Seaside and I lived at 1760 S. Columbia, renting from a former city police chief. Through a college buddy who was bartender at the Embers, behind the Palm Beach Cafe, I became good friends with Alex Haralampus and spent many evenings listening to his fine collection of music on state-of-the-art sound system and playing a game of chess on the bar with Alex. During 1957-58 another one of my favorite dining places was the Partee Room that overlooked the golf course on Avenue U. The golf course was run by the Cartwrights at that time and the chef was an outstanding Philippino cook, Tommy.

The last hotel shows the house plus two apartments, the barn and riding ring. The garages were torn down but held an old six person carriage we used to pull around by hand when we were kids. The walkway to the beach enters into the sun porch. The swimming pool was built in the early 50's. The horses would leave from the stables and head up to Surf Pines through the Dunes and lower Palisades, then return by the beach. SANDY MCCALL

Gearhart Kids in the 1950's, identified as Tom Greenough, Jan Herron, Mark Goodell, Sandy McCall, Preston Devereaux, Kate Osburn, Sally Herron, Blake Osburn, Jim Lucas, and Mike Herron.

"The Cobra Gardens, American Cafe, Tico Taco Mexican, and Oregon Trail Trading Post; World famous live reptile display, Reuben the Great, king cobra; Susie, the 200 pound, twenty foot python; rattle snakes, gila monster, Gearhart Junction Highway 101."

The WRECK of the Peter Iredale

1906

The most photographed
Shipwreck in the country
Is shown at different times
During the past 100 years in
The same place it stopped
Near fort stevens in 1906.

1962

1952

Camp Kiwanalong

Camp Kiwanalong was founded by the Kiwanis Club of Astoria in 1936 as a Girl Scout Camp. The name Kiwanilong was conceived by an eight year old child, meaning Kiwanis on Long Lake.

"On an evening in May, three of us girlfriends decided to go to Cullaby Lake, but when we got there we decided to go swimming. Since we didn't have bathing suits with us, we'd go skinny dipping. After playing in the water we got out to put our clothes back on and found them gone! So we sat in the car and finally decided to go to the caretaker's house and asked to use the phone, which was a little embarrassing since we were without clothes. We were lucky as the wife of the caretaker came to the door. We proceeded to tell her our circumstances and that we needed to phone one of our parents for a little help here. She gave us blankets until a parent arrived and we got home. On the next day at school, we thought someone would say something. We didn't say anything either, and we never heard a word, and we never saw our clothes again."

by Patsy Kerwin

A New Road to Cannon Beach

Ecola Inn

It was very difficult to get to Cannon Beach. There was a winding road to maneuver and from the south side, it was passable only as long as the tide wasn't in over the coast highway at Hug Point. A new highway was constructed in 1952 and was the first successful bypass in the state of Oregon. Cannon Beach had been isolated for many years except for the few families who lived there year around and the ones who stayed in the summers. Out of the ocean with it's huge rocks, a star was born on the Pacific Coast.

Arch Cape 1952

Freak Wave 1962

Angela Baumgartner

Twilight on the Prom — Lisa Johnston

The Flavel House Astoria

This historic 1885 Queen Anne Style Victorian Home and furnishings shows what life was like in Astoria at the turn of the last century.

Lisa Johnston

Columbia River Maritime Museum

The Columbia River Maritime Museum was founded in 1962 when Rolf Klep, a native of Astoria, returned to his birthplace after retirement as a successful graphic artist on the east coast. Upon returning, Klep (a long-time collector of marine artifacts) and a group of his colleagues sought to establish a museum to preserve the rich maritime heritage of the entire Columbia River region - a museum of national distinction. Today the Columbia River Maritime Museum's national reputation for the quality of its exhibits and the scope of its collections makes it one of the finest maritime museums in the nation. It is also the first museum in Oregon to meet national accreditation standards, and has been designated the official state maritime museum for Oregon.

David Rix 2008

North Head Lighthouse

Completed in 1898, the North Head Lighthouse was built in response to a marked increase in the number of shipwrecks along the Peninsula as mariners coming from the north could not see the Cape Disappointment light. The 65-foot North Head Lighthouse was designed by German-born engineer C.W. Leick and sits on solid basalt more than 190 feet above sea level.

Looking South from North Head

Broadway 1971

The Pacific Ocean 1964 E A Weed

Randy Curs

Sunset Lake

The Columbia   Preston Devereaux 2008

Angela Baumgartner is a Pacific Northwest artist whose photography captures people, nature, and brief moments of the beauty that surrounds her.

Credits and Acknowledgements

Jerry Alto
John Alto
Angela Baumgartner
Novella Crafard
Randy Curs
Jeff and Cindy Daly
Preston Devereaux
Russ Earl
Bob Furrer
Mary Hill Cornell
Lisa Johnston
Deborah Harmon Lee
Kevin Kent
Lou Ann Kerr Smith
Carl Larson
Tom Lenocker
Donna Cameron Matteson
Karl Marlantes
Dena and John Marsch
Sandy McCall
Anna Mehelis McNeil
Crayton Morris
Les Morris, Jr
David Rix
Nancy Scott Battilega
Trudi Schnitzer Stone
Jeanette and Bud Stevens

Many thanks to
Donna, Mary and Lou Ann,
For making this one happen.

to
Judy and Jennifer,
for Copy Editorial help.

Thank you
Angela and Lisa,
for your beautiful photography.

And a very special thanks
to my Amazing Friends,
for your stories and enthusiasm!

References

Appelo, Carlton, "Deep River, "The C. Arthur Appelo Story" Wahkiakum County, 1978.

Gibbs, James, "Pacific Graveyard" 1964, Binford & Mort Pub.

Hanson, Inez Stafford "Life on Clatsop" published 1970

Jennings, Harold "Ferry Boats & Dreams" Copyright 1983

Larden, Dennis, "Legacy" Copyright 1974, Windfall, Rick Nelson and the Stone Canyon Band

Miller, Emma Gene, "Clatsop County, Oregon", 1958 Metropolitan Press, Portland, OR

Oregon Historical Society, Portland, Oregon

Oregon State Archives, Salem Oregon

Peterson, David, "Celiast Smith: Pioneer Woman, Indian Woman" Louis Pelzer Award Entry 1992, Univ. of Oregon, Eugene, OR

Penner, Liisa, Clatsop County Historical Society

Penttila, Bryan, "Columbia River, the Astoria Odyssey", Amato Publishing, Inc. 2006.

"The First 75 Years", published 1981, Seaside Signal, Seaside, OR

Thomas, Annalou B., Gearhart Hotel, Cumtux Vol. 22. Spring

US Naval Historical Center, Cross, "The Voyage of the Oregon San Francisco to Santiago in 1898", Boston, Merrymount Press, 1908

Waller, Robert James, "The Ballads of Madison County" Blue Suspenders, Copyright 1993, Atlantic Recording Corp.

Wallingford, Roger, Portland, OR, Celiast Coboway Smith family.

Witt, Sandra, "Deep River Finns" Internet Deep River, WA

Index

SEASIDEBOOK.COM